Oxford
International Primary
Geography

Terry Jennings

2

OXFORD
UNIVERSITY PRESS

Great Clarendon Street, Oxford, OX2 6DP, United Kingdom

Oxford University Press is a department of the University of Oxford. It furthers the University's objective of excellence in research, scholarship, and education by publishing worldwide. Oxford is a registered trade mark of Oxford University Press in the UK and in certain other countries

British Library Cataloguing in Publication Data
Data available

978-0-19-831004-4

10 9 8

Paper used in the production of this book is a natural, recyclable product made from wood grown in sustainable forests. The manufacturing process conforms to the environmental regulations of the country of origin.

Printed in China by Sheck Wah Tong Printing Press Ltd.

Acknowledgements
The publishers would like to thank the following for permissions to use their photographs:

Cover photo: Getty Images/Stephan Studd, P4: Pavel Vakhrushev/Shutterstock, P5a: Stocktrek Images/Getty Images, P5b: MarcelClemens/Shutterstock, P7: E+/Getty Images, P8a: Stockbyte/Getty Images, P8b: Sementer/Shutterstock, P8c: Oleg_Mit/Shutterstock, P9a: Lawrence Manning/Spirit/Corbis/Image Library, P9b: Pashin Georgiy/Shutterstock, P9c: Asaf Eliason/Shutterstock, P11a: Denis Burdin/Shutterstock, P11b: Shutterstock, P12: Lonely Planet Images/ Getty Images, P13a: Anadolu Agency/Getty Images, P13b: E+/Getty Images, P14a: Shutterstock, P14b: keith morris / Alamy, P14c: Volker Steger/Science Photo Library, P15: Dreamstime, P16: The Image Bank/Getty Images, P17a: OPIS Zagreb/Shutterstock, P17b: Yann Arthus-Bertrand/Documentary Value/Corbis/Image Library, P18: The Image Bank/Getty Images, P19: Kevin Phillips/Photographers Direct, P20: Walter Bibikow/Encyclopedia/Corbis/ Image Library, P21a: Falcon Cinefoto/Art Directors & Trip Photo Library, P21b: photo360/Dreamstime, P22a: Rob Howard/Encyclopedia/Corbis, P22b: Pavelk/Shutterstock, P23a: All Canada Photos/Getty Images, P23b: All Canada Photos/Getty Images, P24: Shutterstock, P25a: Frans Lemmens/Eureka/Corbis/Image Library, P25b: hemis.fr/Getty Images, P26a: Katja Kreder/Passage/Corbis/Image Library, P26b: Eniko Balogh/Shutterstock, P27: Konstik / istock, P28a: Eye Ubiquitous/Alamy, P28b: Alexander Kondakov / Alamy, P29a: FLPA / Alamy, P29b: Paul A. Souders/ Encyclopedia/Corbis/Image Library, P30: Kuttelvaserova Stuchelova/Shutterstock, P34: Nasa, P34: Shutterstock, P36: Shutterstock, P37: Chantal de Bruijne/Shutterstock, P38a: Mohammed Huwais / Stringer/Getty Images, P38b: Andre Jenny / Alamy, P40: kelvin tran/Shutterstock, P41: Panorama Media/Gettyimages, P42a: Moment/Getty Images, P42b: Photographer's Choice/Getty Images, P43: Zhu Xudong/Xinhua Press/Corbis/Image Library

Although we have made every effort to trace and contact all copyright holders before publication this has not been possible in all cases. If notified, the publisher will rectify any errors or omissions at the earliest opportunity.

Links to third party websites are provided by Oxford in good faith and for information only. Oxford disclaims any responsibility for the materials contained in any third party website referenced in this work.

Contents

The Earth

Hundreds of years ago people thought the Earth was flat.

Sailors on long sea journeys were afraid they would fall off the edge of the Earth.

Now we know the Earth is round, like a huge ball.

We call this shape a **sphere**.

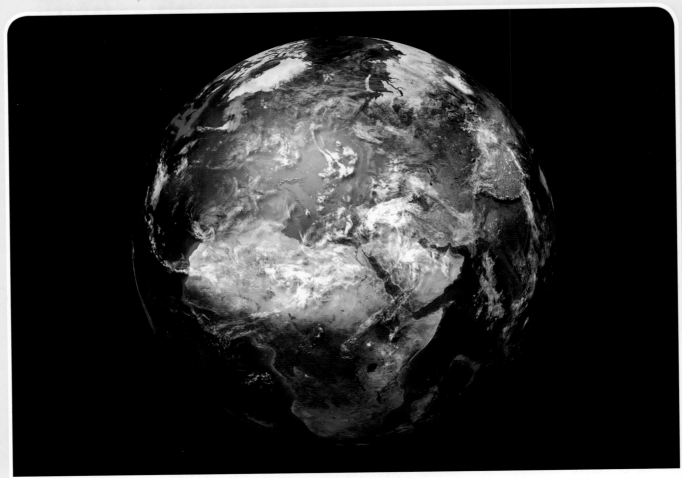

4

What can you see in this picture of the Earth taken from space? What are the white shapes?

The Earth from space

Seen from a spaceship, the Earth looks blue.

This is because there is much more water on the Earth than there is land.

Huge oceans take up most of the Earth.

Air and water

We are lucky that there is water on the Earth because we need water to live.

The Earth also has a thick layer of air around it.

We need the air to breathe.

Without air and water we would not be able to live on the Earth.

We also need the Sun to give us warmth and light and to help plants grow.

Night

Day

The Sun only shines on half the Earth at a time. Where the Earth is facing the Sun it is day time. Where the Earth is not facing the Sun it is night time.

Activities

Use a blank outline **map** of the world. Colour the land and the sea. Use crayons or felt-tipped pens for this. Add a key to show what your colours mean.

Continents

There are seven huge areas of land on the Earth.

They are called **continents**.

Seven continents

The continents are Europe, Asia, North America, South America, Africa, Oceania and Antarctica.

Some continents are joined to each other.

Others have water all around them.

Which of the continents have water all around them?

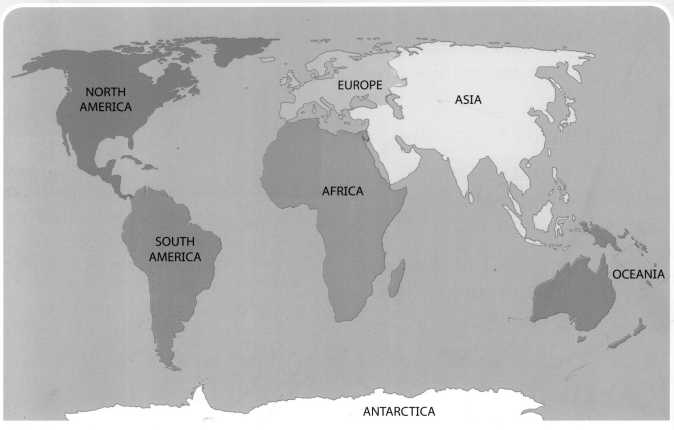

The seven continents.

Asia

The largest continent is Asia.

It covers one third of the Earth's surface.

More people live in Asia than in any other continent.

More than half of the people in the world live in Asia.

Antarctica

The continent with the fewest people is Antarctica.

The only people who live there are scientists.

They are there to study the world's coldest continent.

Scientists in Antarctica live and work in buildings like these.

Activities

Choose one of the continents. Find out about its:

- countries
- main **mountains**
- main rivers.

Tell the rest of the class what you have discovered.

Countries

What is a country?

A country is an area of land that has its own name, its own **government** and its own flag.

Most countries also have their own money and their own language.

The money and flag of Pakistan.

Altogether there are more than 190 countries in the world.

All of the **continents**, except for Antarctica, are divided into countries.

Europe is made up of 46 countries, while there are 54 countries in Africa.

South America has only 12 countries.

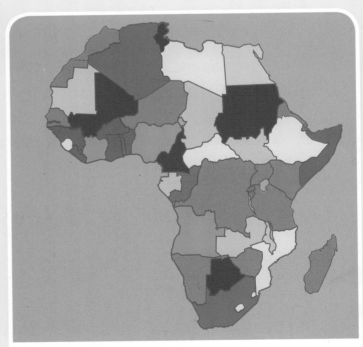

How many countries can you see on this **map** of Africa?

Countries large and small

The biggest country in the world is Russia.

It takes nearly a week to cross Russia by train.

The smallest country is Vatican City in Rome.

You can walk across it in one hour.

The money and flag of Russia, the world's biggest country.

Activities

1 A **border** is the line that marks the edge of a country. Some borders are wiggly; some borders are straight.

   Look at a map or **atlas** and list five countries with straight borders.

2 Some countries have the sea as a border. They are countries with a **coastline**.

   Using your map, write the names of six countries that do not have a coastline.

Oceans and seas

The huge areas of water on the surface of the Earth are called oceans.

The oceans cover nearly three quarters of the Earth's surface.

There are five separate oceans: the Pacific, the Atlantic, the Indian, the Arctic and the Southern or Antarctic Ocean.

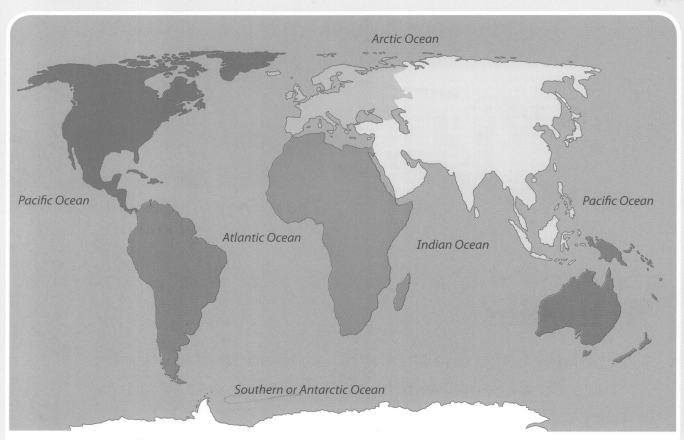

Arctic Ocean

Pacific Ocean

Pacific Ocean

Atlantic Ocean

Indian Ocean

Southern or Antarctic Ocean

The world's five oceans.

The Pacific Ocean

The Pacific is the largest and deepest of the world's oceans.

It is bigger than all the Earth's land put together.

The Arctic Ocean

Most of the Arctic Ocean is frozen and covered with ice.

In the summer some of the ice melts. Huge blocks of ice, called **icebergs**, float away.

Icebergs in the Arctic Ocean.

Mountains, valleys and volcanoes

The bottom of the oceans is not flat.

There are high **mountains**, deep **valleys** and even **volcanoes** on the ocean floor.

Seas

Within the oceans are smaller areas of water called seas.

They are cut off from the open oceans by pieces of land.

The water in the Dead Sea in Jordan is so salty you can float in it.

Activities

1 Make a class wall display about oceans. Use old newspapers and magazines for pictures and information.

2 Use an **atlas** and find:

   a the name of the ocean that lies between Europe and North America.

   b the name of the ocean between the North Pole and Greenland.

   c the name of the ocean between Africa and Australia.

My environment

The world around us is called our environment.

Our environment includes the air we breathe, the water we use and the land around us.

It also includes the places where we live.

Plants and animals are part of our environment, too.

Our environment is everything around us.

Using the environment

Everything we use comes from our environment.

Our food and water come from our environment.

Our clothes, homes and toys are all made from materials from our environment.

Damage to our environment

Even small changes can damage our environment.

Dropping litter or making a lot of noise can spoil the environment for other people.

Wasting water and electricity is also bad for our environment.

Towns, cities and the countryside can all be good to live in if we look after them.

In some cities the air is so dirty that people wear special masks.

All living things, including people, need clean water if they are to stay healthy.

Activities

1 a Work with a group of friends. Discuss what you like best about your environment.

   b What don't you like about it? What do you think could be made better?

2 Use a box, such as an old cereal packet, to make a small litter bin for your table or the library corner.

Recycling at home

Every day we produce a lot of waste objects and materials.

Recycling

Many of the things we throw away can be reused.

Many of them can be turned into something useful.

This is good for our environment.

Some things are **recycled** already.

Why is a rubbish tip like this a waste of valuable materials? Why is it bad for people's health?

Many cities, towns and villages have recycling centres to recycle glass, metal, paper and clothes.

Glass bottles and jars that are recycled can be crushed and melted and then made into new bottles and jars.

Is there a recycling centre near your home or school? Do you use it?

This recycled glass has been melted and will be made into new glass containers.

Compost, cans and paper

Fruit and vegetable peelings can be put onto a compost bin to rot away.

Old metal cans can be melted down and used to make new metal objects.

Old magazines and newspapers can be used to make new paper.

Protecting our environment

If we recycle old paper, fewer trees are cut down. If we recycle glass and metals, fewer materials that come from the ground are needed. Then there are fewer big holes in the ground.

Old fruit and vegetable peelings can be put into a compost bin to rot away. The compost can then be put onto the soil to help more plants to grow.

Activities

1 Draw a picture or diagram on a large sheet of paper showing what happens to a bottle or jar that you have used at home and then taken to the recycling centre.

2 Design a poster to encourage people to recycle their old bottles, cans, paper, plastic, clothes and other waste materials.

② An island home

Islands

An island is a piece of land with water all around it.

Most islands are separated from the **mainland** by the sea.

A group of small islands seen from an aircraft.

Large and small islands

Look at the **map** on pages 44 and 45.

How many islands can you find?

The world's largest island is Greenland.

There are many other islands.

Australia is one huge island continent, while the Maldives is made up of 1200 small islands.

Near islands

Some islands were once joined to the mainland.

Bahrain and Singapore were once joined to the mainland.

Thousands of years ago, they were cut off by the sea.

The sea has cut off a piece of land to make an island.

Distant islands

Some islands lie far out to sea.

The Hawaiian Islands are the tops of high **mountains** on the bottom of the sea.

Iceland and Tahiti were made by **volcanoes** under the sea.

In warmer seas, many islands are made up of the shells of tiny sea animals.

These are called **coral** islands.

A coral island in the Pacific Ocean.

Activities

a Look at a world map or a **globe**. Choose an island. Write a sentence or two saying where it is and what it is like.

b Use reference books or the Internet to find out more about your chosen island. Share your work with your class.

Bahrain, an island country

Bahrain is a small country in the Arabian Gulf.

It is a group of 33 islands.

On the islands

More than a million people live on the islands of Bahrain.

Most of them live in or near the **capital** city, which is called Manama.

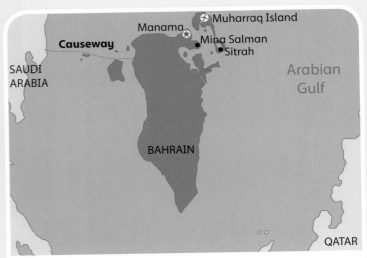

The main islands that make up the country of Bahrain.

Modern buildings in Manama, the capital of Bahrain.

Making money

The money used in Bahrain is called the Bahrain dinar.

Bahrain earns most of its money by selling oil and natural gas to other countries.

Some of Bahrain's money also comes from people who visit Bahrain on holiday.

Farming

Most of Bahrain is low **desert**.

The farmers on Bahrain grow different kinds of fruits and vegetables.

Dates are grown around the **oases**.

Some farmers also keep chickens and other animals.

Bahrain cannot produce all the food it needs so a lot has to be bought from other countries.

Farmers growing vegetables in Bahrain.

Activities

1 **a** Do you think Bahrain would be a nice place to live and work? Tell a friend what you think.

   b What job would you like to do there? Say why.

2 Make a poster advertising Bahrain as a good place to visit for a holiday.

Island transport

Many islands are a long way from the **mainland**.

They can only be reached by boat or aircraft.

The islands of Bahrain are near to each other and the mainland.

It is easy to reach them by road.

The roads cross the water on **causeways**.

The longest causeway is 24 kilometres long.

It links Bahrain to Saudi Arabia.

This causeway is 24 kilometres long. It links Bahrain to Saudi Arabia.

Ports

Bahrain has three big **ports**.

Between them, they handle container ships, cruise ships and oil tankers.

A **ferry** sails from the port of Mina Salman to Bushehr in Iran.

Oil tankers loading at the port of Sitrah.

Air travel

Bahrain Airport is on Muharraq Island.

More than 40 airlines have flights to cities all over the world.

Bahrain's busy international airport.

Activities

1 Draw or paint a picture of Bahrain, showing the sea. Add a crane, a container ship, some **cliffs** and **beaches**, or some of the other things you might see on the island.

2 There are good hospitals and doctors on Bahrain. This is not so on many small islands that are a long way from the mainland. Discuss with a friend what you would do if you were very ill or badly injured on a small island a long way from the mainland.

3 Going to the seaside

At the seaside

What is the seaside?

The **seaside** is where the land meets the sea.

At the edge of the sea is the **beach**.

Some beaches are made of **sand**, some are made of small stones called shingle. Some beaches have **cliffs** behind them.

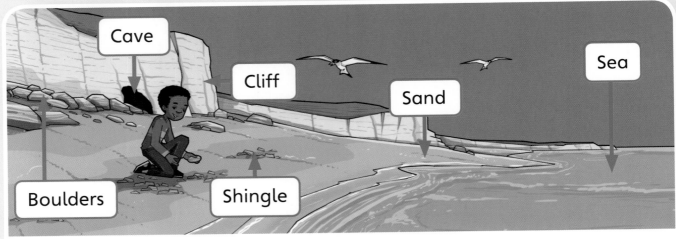

Cave

Cliff

Sea

Sand

Boulders

Shingle

What is the beach made of in this picture?

There are also seaside places where the beach is rocky or muddy.

This beach is made of mud.

This beach is covered in rocks.

Seawater

Have you ever tasted seawater?

The water in the sea is always salty.

This salt is washed into the sea from rocks on land.

Seawater is always moving because of **waves** and **tides**.

Waves are made by the wind blowing across the water.

Tides are caused mainly by the Moon.

When the sea covers the beach, we say the tide is 'in' or it is 'high tide'.

At other times we say the tide is 'out' or it is 'low tide'.

High tide.

Low tide.

Activities

1 Use a **map** or an **atlas** to find the seaside place nearest to your home.

   a What is it called? Is it a village, a town or a city?

   b What kind of beach does it have?

   c Roughly how far away from your home is it?

   d Try to find pictures of the seaside place. Tell a friend what it is like.

2 Look at the pictures of seaside places on these two pages. Which would you most like to visit? Say why.

Seaside cities, towns and villages

Look at an **atlas** or **map** of the world.

How many towns and cities can you find that are near the sea?

There are also thousands of villages near the sea.

Seaside villages

Many **seaside** villages were built on a hill or **cliff** near the sea.

These places were easy to defend against enemies.

Other seaside villages were built where there were sheltered **harbours** or where there was good fishing.

Some of these early villages grew into towns and cities.

This small village in Italy was built on a cliff overlooking the sea.

The importance of rivers

Some towns and cities are situated where a river meets the sea.

This is a good place to build a **port**.

Large ships, which carry **cargoes** from one country to another, can stop there.

Some of the biggest cities in the world grew from villages that were built near the mouths of rivers.

Holiday resorts

Some towns have developed because people go there for holidays.

These are called holiday **resorts**.

Resorts have places where holidaymakers can stay and fun things for them to see and do.

Rotterdam, one of the world's largest ports, is at the mouth of the River Rhine in the Netherlands.

Sharm El Sheikh is a popular seaside resort on Egypt's Red Sea **coast**.

Activities

1 On a blank outline map of the world, mark and label some cities and resorts that are on the coast.

2 **a** Carry out a survey. Ask your friends and relatives:

- Where is their favourite holiday resort?
- Is it a city, a town or a village?
- Why do they like it best?

b Draw a bar chart of your findings and show them to the class.

Seasides around the world

Nowadays, there are thousands of **seaside resorts** all around the world.

Malta

Malta is a small island country in the Mediterranean Sea.

Hot, dry summers and warm, sunny winters bring many holidaymakers to Malta.

There are many hotels and there are boat trips so that holidaymakers can see the beautiful scenery.

A seaside resort on the island of Malta.

Florida

Florida is part of the United States of America.

The **weather** is always warm and sunny so people can go for holidays at any time of year.

Florida has fine **sandy beaches**.

It also has many places to see and things to do.

The Space Center at Cape Canaveral is in Florida.

Jamaica

Jamaica is an island in the Caribbean Sea.

The weather is warm and sunny all the year round.

Behind many of the beaches are large hotels.

Every year more than a million holidaymakers visit Jamaica for its warm sunshine, beautiful beaches and scenery.

Jamaica has many fine sandy beaches.

Activities

1 Use an **atlas** to find Malta, Florida and Jamaica.

   a If you were to travel to each of these places from your home, how would you get there?

   b Which countries, oceans and seas would you cross to reach each of these places?

2 Work with some friends. Collect some holiday brochures. Find six seaside resorts in different parts of the world. Discuss which resort you like most. Write a few sentences to say why you would like to spend a holiday there.

Food from the sea

26-11-2018

Fish are an important food.

Most of the fish we eat come from the oceans and seas.

Fish traps

Some fish are caught in a trap like this.

The trap contains food for the fish to eat.

When the fish go into the trap to eat the food, they cannot get out.

Long nets

Some fish are caught by fishing from the **beach**.

A long net is dragged through the water.

The net surrounds the fish which are then pulled ashore.

The trawler

A **trawler** goes to sea for days or weeks.

It drags a net like a huge string bag through the water until it is full of fish.

The fish are frozen to keep them fresh until the trawler gets back to **port**.

Fish farming

Fishing the oceans and seas can be difficult and dangerous.

Now scientists are trying new ways of farming fish in the sea, for example using large cages like these.

Activities W.H

1 Find out what kinds of fish and shellfish are sold in your local supermarket. Try to find out where the fish or shellfish were caught. This may be written on the packet or you could ask the fishmonger. Can you find these places on a **map** or **atlas**?

2 Collect postcards and photographs from magazines that show the different methods of fishing. Make a class display of your pictures.

Making sense of the world

How do we learn about the world around us?

Our **senses** tell us what is happening.

What are the five main senses?

Working senses

Our senses are working all the time.

What can you see now?

What can you hear?

How can you tell if it is hot or cold?

Why does the smell of food make you feel hungry?

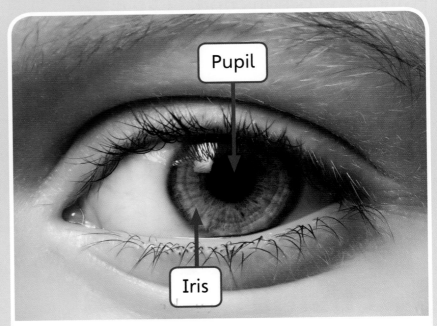

The coloured part of the eye is called the iris. Light goes into the eye through a hole called the pupil.

Being observant

We see with our eyes.

Sight is an important sense.

If you really look hard, you can see things you have never noticed before.

This is called being observant.

30

How observant are you?

There are ten differences between these two pictures. Can you spot them all?

Activities

1 Draw and label something that you can: see; hear; feel; smell; taste.

2 How can your eyes help you to keep safe? Write down five different ways.

Seeing small or distant objects

We can learn many things about the Earth with the help of **lenses**.

Binoculars make things that are far away seem closer and larger. When would you use binoculars like these?

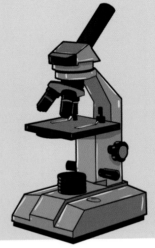

A microscope lets us see tiny things that are too small to see with only our eyes. When would you use a microscope?

A magnifying glass makes things look bigger.

A telescope helps us to see things that are far away.

Projectors, microscopes, binoculars and telescopes all have lenses.

They help us to see things that are too small or too far away to be seen with our eyes alone.

Lenses in cameras, video cameras and camcorders let us make pictures of the things that we see.

Artificial satellites

Artificial satellites also collect information about the Earth.

They are sent into space by rockets and they circle the Earth.

These are just some of the things satellites are used for:

- sending television and radio programmes and telephone messages around the world
- helping people in cars, ships and aircraft to find their way
- collecting information about the **weather**
- studying the rocks, soils and plants on the Earth's surface
- making **maps**.

Activities

1 Make a collection of different things with lenses in them.
   - Which lenses make things look bigger?
   - Which lenses make things look smaller?
   - Which lens is best for looking at something far away?
2 Watch the weather forecasts on television. Can you see any photographs taken from satellites?

This is the Earth, seen from space.

Which **continents** can you see?

Globes

A globe is a ball with a **map** of the world on it.

It shows the shape of the land and oceans.

Globes are hard to carry around.

A globe cannot be folded up and put in your pocket, but a map can be.

A globe.

Maps

A map is a flat picture of the Earth seen from high above.

A map can show all the Earth's surface at the same time, stretched out on a page.

Why can't a globe do this?

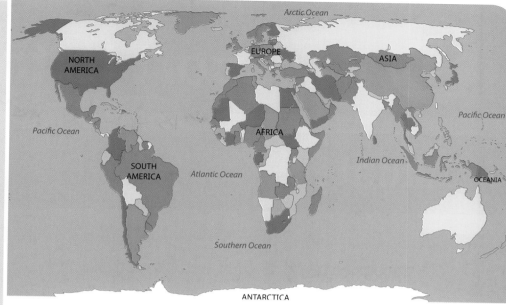

A map of the Earth. You can see all the continents, oceans and seas.

Atlases

An **atlas** is simply a book of maps. Some atlases show all the places on Earth. Others show only the roads in a particular country or area.

Activities

1 a Lay some tracing paper on a globe. Carefully draw round one of the continents.

   b Now lay your tracing next to the same continent on a map of the world. How are they the same? How are they different? Why are they different?

2 Look at the labels in your clothes. Use an atlas to find the countries where your clothes were made. Write a sentence saying what each item is and where it came from.

18-10-2018

Climate

The **climate** of a place is its usual **weather** over many years.

Different parts of the world have different climates.

The **Equator** is an imaginary line around the middle of the Earth.

Near the Equator, the Sun is high in the sky all through the year.

The Sun's rays are very strong and hot.

The Equator passes through Brazil, Colombia and Kenya.

All of these countries have hot weather every day.

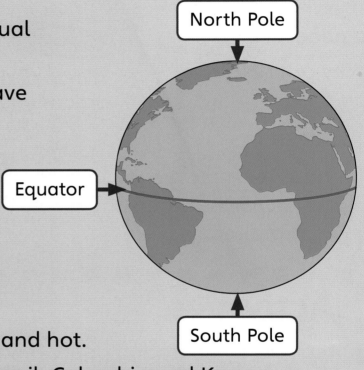

North Pole

Equator

South Pole

The Poles

Around the North and South Poles it is very cold all the time.

There is always a lot of ice and snow.

Near the South Pole in Antarctica it is very cold all year.

This is because the Sun is never very high in the sky.

Its rays are weak and not very warm.

Near the Equator it is hot all the year round, although not everywhere is as dry as this.

In-between climates

Away from the Equator and the North and South Poles, it is usually neither very hot nor very cold.

It is warm in summer but colder in winter.

The United Kingdom and New Zealand have a climate like this.

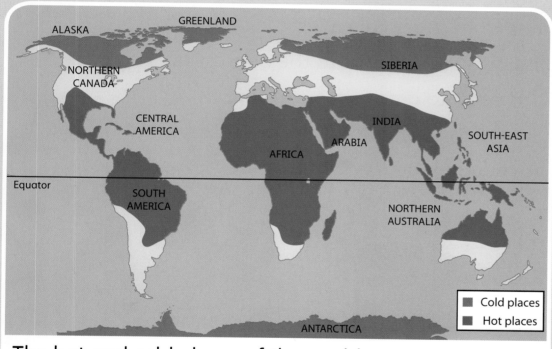

The hot and cold places of the world.

Activities

1 Look at an **atlas** or a **globe** and make a list of as many countries as you can through which the Equator passes.

2 Can you find two large islands on the Equator? What are they?

The global supermarket

30-10-2018

Food helps us to grow and stay healthy.

It also gives us the energy to walk, talk, run, sleep and do all the other things we do.

Some foods come from animals, some come from plants.

Growing and buying food

Some of our food is grown in our own country.

Some of it comes from other countries.

After the food is packed, it is sent to the nearest **port** by truck.

A ship carries the food to a port in our country.

Trucks then take the food to our local shop or supermarket.

Most people have to buy their food from shops, markets, supermarkets and other places.

The labels on food tell us what is inside each container. They often also tell us the name of the town or country where the food was grown or packed.

These bananas were picked on an island in the Caribbean Sea. The boat will take them to a large ship.

A food map

Look at this food **map**.

Each food label is joined to the country the food comes from.

Which of the foods would have to travel furthest to reach you?

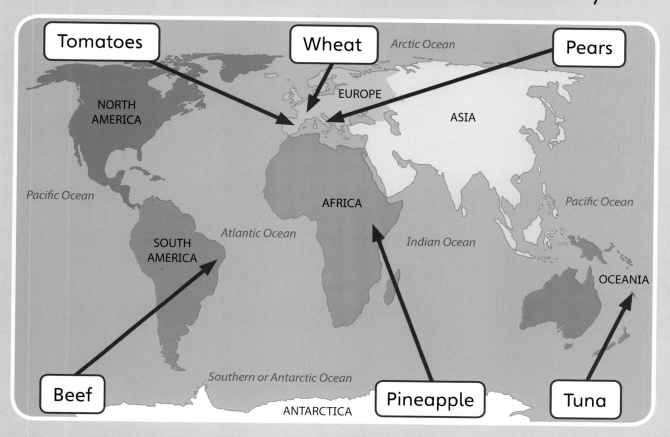

Activities

1 Make a food map, like the one shown, for a class display.

2 Make a class collection of pictures of different foods. Make sets of your pictures. You could have:

- a set of foods which come from plants and a set of foods which come from animals
- a set of foods grown or made in your country and a set which come from other countries
- a set of fresh foods and a set of frozen or dried food.

China

Find China on the **map** of the world on pages 44 and 45.

China is one of the largest countries in the world.

Only Russia, Canada and the United States of America are larger than China.

More than 1.3 billion (1300 million) people live in China.

That is more than in any other country.

Where people live

China has some of the highest **mountains** in the world.

There are also huge **deserts**.

Most people live near the great rivers that flow across China.

Many of them are farmers who grow rice and other crops.

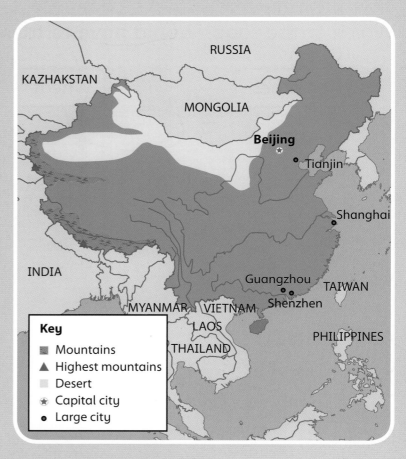

Many of the farmers in China grow rice in flooded fields like these.

Chinese cities

Look at the map of China on page 40.

What is the **capital** city of China?

Which other large cities can you see?

China is changing

China used to be a poor country, but that is changing.

China is building new schools, farms and factories.

China has now become the world's biggest seller of goods to other countries.

About 25 years ago, very few Chinese families had a car.

They used bicycles to get to work or school, or to the shops.

Now millions of Chinese families have a car.

Many families in China now have a car, making the streets in the cities very busy.

Activities

I Look at a map of the world.
If you were to fly from your home to Beijing, which countries, seas and mountains might you fly over? Find something out about each of the countries you might fly over.

2 a Make a class collection of pictures and small things that show what life is like in China.

b Label each picture and thing.

c Say what it shows about China.

Living in Shanghai

Look at this picture of Shanghai at night.

How is it different from where you live?

Shanghai is the biggest **port** in China and also the largest port in the world.

The city is built beside the Huangpu River.

Shanghai is one of the largest and busiest cities in the world.

Meet Mei-Ling

Mei-Ling lives with her parents in a small apartment.

She is 6 years old.

Like most children in China, Mei-Ling is an only child.

She has no brothers or sisters.

A school day

For breakfast, Mei-Ling eats rice porridge with vegetables and steamed bread.

Lunch is usually noodles with stir-fried vegetables and perhaps a little meat or fish.

Mei-Ling walks to the local primary school.

School starts at 7 a.m. and doesn't finish until 4.30 p.m.

Lessons

At school, Mei-Ling's class has more than 50 children.

They study Chinese, mathematics, music, craft, physical education and science.

Mei-Ling's favourite lesson is craft.

Mei-Ling's school uniform is a red tracksuit.

Activities

1 Copy and complete this table to compare your life with Mei-Ling's life.

	Mei-Ling	Me
Age	6 years	
Home	Small apartment	
Brothers and sisters		
Breakfast food		
School uniform		
Number of children in class		
Favourite lesson		

2 Would you like to live in Shanghai? Say why.

Map of the World

GREENLAND

ICELAND

UNITED
KINGDOM

NETHER

CANADA

UNITED STATES

HAWAIIAN
ISLANDS

MEXICO

JAMAICA

COLUMBIA

BRAZIL

CHILE

ARGENTINA

RUSSIA

IRAN

BAHRAIN

QATAR

EGYPT

SAUDI
ARABIA

UNITED ARAB
EMIRATES

PAKISTAN

CHINA

INDIA

MOCRATIC
EPUBLIC
CONGO

KENYA

Singapore

AUSTRALIA

NEW
ZEALAND

NTARCTICA

Glossary

Artificial satellite A machine sent into space to help us communicate with one another or to collect information about the Earth.

Atlas A book of maps.

Beach The strip of sand, shingle, mud or rock where an ocean, sea or lake meets the land.

Border A line that marks the edge of a country.

Capital The most important city in a country.

Cargo A load of goods carried by a truck, train, aircraft or ship.

Causeway A raised road or walkway over water.

Cliff A steep wall of rock, especially on the coast.

Climate The typical weather of a place over a whole year.

Coast The seashore and the land close to it.

Coastline The line on a map marking where the sea meets the land.

Continent One of the seven big pieces of land in the world.

Coral A hard substance made from the shells of tiny sea animals.

Desert A large area of land where few plants can grow because it is either too dry or too cold.

Equator A line drawn on maps to show places half-way between the North Pole and the South Pole.

Ferry A ship used for carrying people or things across a river or narrow sea.

Globe A ball with a map of the whole world on it.

Government The group of people who are in charge of a country.

Harbour An inlet of the sea which gives ships a place to unload or shelter from bad weather.

Iceberg A very large block of ice floating in the sea.

Lens A curved piece of glass or plastic used to make things look larger or smaller.

Mainland The main part of a country, not the islands around it.

Map A drawing of part or all of the Earth's surface as if you were looking down on it.

Mountain A very high part of the Earth's surface.

Port A harbour or a town or city with a harbour.

Recycle To treat waste material so that it can be used again.

Resort A place where people go for their holidays.

Sand The tiny grains of rock that you find on beaches and in deserts.

Seaside A place, such as a village, town or city, by the sea.

Sense The ability to see, hear, touch, taste or smell.

Sphere A globe; the shape of a ball.

Tide The rising and falling of the level of the sea, which happens twice a day. This is caused mainly by the Moon's gravity pulling the water on the Earth.

Valley A line of low land between hills or mountains.

Volcano A mountain with a hole at the top through which hot, molten rock comes out from deep inside the Earth.

Wave A moving ridge of water on the sea, formed by the wind blowing over the water.

Weather The rain, wind, snow and sunshine, for example, at a particular time or place.